1 Architectural Concept

- Feng Shui and Architecture
- Rites and Architecture
- Symbolism and Architecture
- Dragon Culture and Architecture

2 Architectural Element

- Roof
- Gates and Doors
- Window
- Ridge Ornaments
- Dougongs
- Stylobates
- Traditional Chinese Furniture
- Glazed Tiles on Buildings
- Baofu Colored Patterns in Southern China

3 Imperial Architecture

- Palace Museum in Beijing
- Imperial Palace of the Qing Dynasty in Shenyang

4 Ritual Architecture

- The Temple of Heaven in Beijing
- Dai Temple of Mount Tai
- Beizhen Temple at Mount Lv
- Temple of Lord Guan in Dongshan
- The Confucian Temple
- Longmu Ancestral Temple
- Temple of Lord Guan in Xiezhou
- Temple of the South Sea God in Guangzhou
- Huizhou's Ancestral Temples

5 Religious Architecture

- Buddhist Temples at Mount Putuo
- Three Major Taoist Temples at Jiangling
- Wudang Taoist Palaces and Abbeys
- The Temples of Mount Jiuhua
- Tianlongshan Grottoes
- Yungang Grottoes
- Temple of Tibetan Buddhism in Tongren of Qinghai
- The Eight Outlying Temples of Chengde
- Old Chongfu Temple in Shuozhou
- Datong Huayan Temple
- Buddhist Temples in Jinyang
- North Alp Mount Hengshan and Xuankong Temple
- Jinci Temple in Taiyuan
- Buddhist Temples and Pagodas of Dai People in Yunnan Provin
- Buddhist Pagodas and *Tasha*
- Qutan Temple in Qinghai Province
- Buddhist and Taoist Temples in Qianshan
- Tibetan Buddhist Dagobas and Architectural Decoration
- Kaiyuan Temple in Quanzhou
- Guangzhou Guangxiao Temple
- Foguang Temple of Mount Wutai
- Xiantong Temple of Mount Wutai

6 Ancient Cities and Towns

- Ancient Chinese Cities
- Ganzhou: City of Song Heritage
- Ancient City of Pingyao
- Ancient City of Fenghuang
- Ancient City of Changshu
- Ancient City of Quanzhou
- Architecture of Yuezhong
- Penglai Water City
- Castles to Fight Against Foreign Pirates of the Ming Dynasty
- Zhaojia Castle
- Zhouzhuang
- Gulangyu Islet
- Nianbadu - Ancient Town of Southwestern Zhejiang

7 Ancient Villages

- Xinye Village in Zhejiang Province
- Caishiji
- Buildings in Dong Villages
- Vernacular Villages in Huizhou
- DangjiaVillage in Hancheng
- TangmoVillage - Village of Waterside Street
- Donghuali Walking Street in Foshan
- A Military Village - Zhangbi Village
- Luoshui Village - Realm of Matriarchy on the Shore of Lugu L

8 Folk Architecture

- Quadrangles in Beijing
- Folk Architecture in Suzhou
- Vernacular Dwellings of Yixian County
- Weiwu House in South Jiangxi
- Folk Architecture of the Bai People in Dali
- Naxi Civilian Residences in Lijiang
- The Shikumen Dwellings
- Folk Architecture in Kashgar
- Most Elaborate Structure of Fujian Tulou - Eryi Lou in Hua'an

9 Burial Structures

- The Ming Tombs
- Eastern Tombs of the Qing Dynasty
- Three Tombs of the Qing Dynasty Outside the Shanhai Pass

10 Gardens

- Imperial Gardens
- Chengde Mountain Resort
- Scholars' Gardens
- Gardens in Lingnan Area
- Gardening and Rockeries
- Master of the Nets Garden
- Mo's Manor in Pinghu

11 Academy of Classical Learning and Guildhalls

- The Architecture of Classical Colleges
- Yuelu Academy of Classical Learning
- Three Major Academies of Classical Learning in Jiangxi
- The Chen Clan Academy
- Xiling Society of Seal Arts
- The Architecture of Guildhalls

12 Others

- Towers and Pavilions
- Pagoda
- Ancient Pagodas in Anhui
- The Wooden Pagoda in Yingxian County
- Chinese Pavilions
- Bridges in Fujian
- Stone Bridges in Shaoxing
- Memorial Archways

The Series of 100 Gems of Chinese Architecture

ANCIENT TOWN OF FENGHUANG

Text & Photo by Wei Yili

CHINA ARCHITECTURE & BUILDING PRESS

Publisher's Note

China is a great civilization, and boasts vast territories, abundant resources and time-honored history. Since entering into the new century, she has been attracting more and more attention, and presenting her past glamour and brilliance to the world. The booming economy of modern China and the cultural gem of ancient China have already become topics people are keen to understand and study in an in-depth way.

In the process of over 60 years since its founding, China Architecture & Building Press, a national science publishing institution, has been devoted to promoting and disseminating the outstanding architectural culture of the Chinese nation, pushing the progress of Chinese architectural technology, and presenting China's achievements in construction in all ages to the world. Shouldering the mission of "carrying forward the fine traditional Chinese culture, and strengthening the international influence of the Chinese culture," China Architecture & Building Press has, since the 1980s, paid great importance to communication and cooperation with our counterparts at home and abroad on architectural culture, planned, compiled and published a series of academic books and albums about traditional Chinese architecture, bringing about a wide influence on the public.

The Series of 100 Gems of Chinese Architecture is completed as a result of over 100 Chinese experts and scholars' careful survey and concentrated study on representative ancient and traditional buildings with historical significance throughout China. After years of systematic and scientific review, the series of books is compiled according to such themes and categories as architectural concept, architectural element, palatial architecture, ritual architecture, religious architecture, historic towns and villages, folk architecture, burial structures, gardens, academies of classical learning, guildhalls and other buildings. These books are divided into different volumes based on their themes, and contain rich contents about the historical background, architectural style, architectural features, and architectural culture, attached with excellent photos and line drawings. The series consists of 100 volumes, about 2,000,000 Chinese characters and over 6,000 photos and drawings.

The books boast concise descriptions, a wealth of photos and drawings, and elegant design. Easy-to-understand and handy, they are popular books incorporating both professional and cultural elements and suitable for readers from home and abroad. They will provide those who love Chinese culture with an opportunity to appreciate and understand the unique style of traditional Chinese architecture, characteristic design methods, exquisite architectural skills, and ingenious processing of details, record the remarkable heritage buildings for the world architectural community, and open a window for domestic and overseas readers to access the architectural knowledge and art.

The series will be released in both Chinese and English versions, and will be a valuable resource for researchers, enthusiasts and travelers from home and abroad who are interested in historic buildings.

Contents

011   I. Ancient Town of Fenghuang

021   II. A Town in Landscape

029   III. Existing Vernacular Antiquities

035   IV. Places for Farewell ——Pavilions

039   V. Temples and Ancestral Halls

053   VI. Stone Paved Ancient Streets

063   VII. Scenic Spots Along Tuojiang River

079   VIII. Courtyard Dwellings and Stilted Buildings

092   Chronology of Significant Events

Ancient Town of Fenghuang

Fenghuang County is located in Xiangxi (western Hunan Province) which geographically covers a large area and takes the counties alongside the Yuanshui River as its major part including Jishou, Sangzhi, Dayong, Longshan, Yongshun, Baojing, Guzhang, Huayuan, Luxi, and Fenghuang County in northwestern Hunan Province. Xiangxi is the place where the ancient barbarians from Yunmeng lakes (the Dongting Lake today) retreated to when they were defeated by the Han people in war. Qu Yuan, a poet, in Chu Kingdom during the Warring States Period (475 B.C.—221B.C.), was once exiled and went upstream on boat along Yuan River. Today, most places he visited then still can be found in Xiangxi. The spirits in valleys and caves, stink weed and fragrant flower in his poetry can be seen anywhere in the region. The ceremony that thanks the gods with sacrifices which prevails under the cultural dominance of *Chuci* (The Songs of the South) shares the similarity with Great Nuo exorcise ceremony hosted by Miao witches nowadays in Fenghuang County. Thus, it can be said that the culture of Xiangxi may belong to the Witchcraft Culture system of Chu region.

Since the military forces were dispatched to Miao region in 1795 (the 60th year of Emperor Qianlong's reign), the military system was rearranged and more road were built leading to anywhere in the region. According topography in all counties, roadblocks and military outposts were set up and fortifications were constructed constituting an intricate military network. The guards selected from local strong men for fighting and farming wouldn't expose themselves to the outside, so most of them were just kind as common people. To achieve the goal of "countering Miao with Miao", the ruler created new positions called Miaobian (soldiers of the Miao) for officers from different villages to supervise each other. During the Republic of China, local armies and governmental troops formed several military forces. In the past, to oppose oppression, some brave Miao people entered into mountainous areas to rob the rich and assist the poor, which was the specialty of the society in Xiangxi area. Therefore, it seemed as a barren land suffered most from banditry in strangers' eyes.

In the Spring and Autumn Period (770BC-476BC), Fenghuang County was a part of ancient Chu Kingdom. Its organizational system has undergone changes through the vicissitudes of dynasties. In 1707 (the 46th year of Emperor Kangxi's reign), chieftain system was abolished and local governance of sub-prefecture and county is implemented with "Daotai" (Intendant of Circuit) and "Zhengtai" (The Chief General) as their chief respectively. In the period of the Republic of China, Fenghuang Ting was changed into Fenghuang County, and then it has been part of Tujia and Miao Autonomous Prefecture of Xiangxi since in 1957.

Fenghuang Town (also called Tuojiang Town) is surrounded by mountains and lies in lower-lying area with beautiful natural scenery. It is a county seat and covers an area of six square kilometers, with a population of 16,000 in 1988. From ancient times, it has attracted lots of visitors from afar, of whom many poets, writers and painters have thought highly of it in their works. For example, New Zealand writer Rewi Alley (1897.12.2-1987.12.27) once said, "There are two beautiful towns in China: one is Fenghuang Town in Hunan; the other is Changting Town in Fujian." The American scholar Jeffrey C. Kinkley (1948-) who studies the works of Shen Congwen said, "I have been to many places in China, which impressed me most was Fenghuang County." This town is not only an area of scenic beauty but also the birthplace of great talents, such as Tian Xingru, the famous general in the Qing Dynasty, Xiong Xiling, the first elected prime minister of the Republic of China, Tian Xingliu, the senior adviser of Sun Yat-sen Mansion, Shen Congwen, a literary giant and, Huang Yongyu, a master of the traditional Chinese painting. Undoubtedly, Fenghuang is a veritable county of "phoenixes" representing talents and gurus.

In 2002, the ancient Town of Fenghuang was approved as a national historical and cultural famous city.

0-1 Fenghuang Region

It is from Volume One of the *Fenghuang Prefecture Records*. There are about 500 forts and 200 military camps surrounding the remote town, as if drums playing and war torches warning at that time were still sounding and visible.

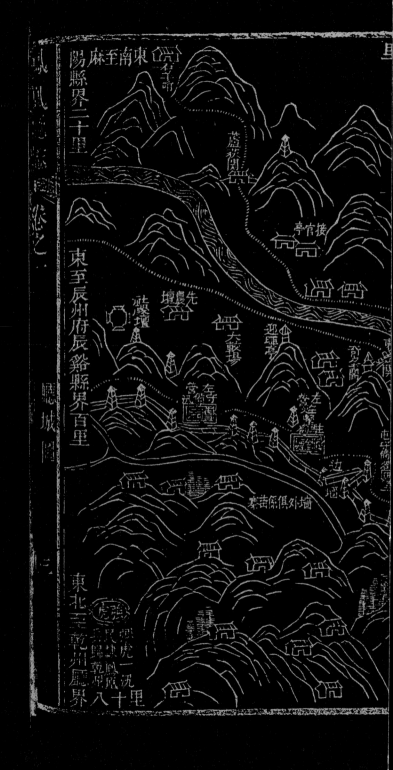

廳境圖

0-2 Surrounding of the Town

It is from Volume One of the *Fenghuang Prefecture Records*, which is an overview of the old town. The town walls, moat, barbican entrance and small drill ground on the north bank of the Tuojiang River are clear on the picture. The stone walls form a town of irregular shape, with five town gate towers standing majestically on their sites. Inside the town, the roads, temples and halls are arranged properly. While dotted nunneries and temples outside the town echo with the mountains around.

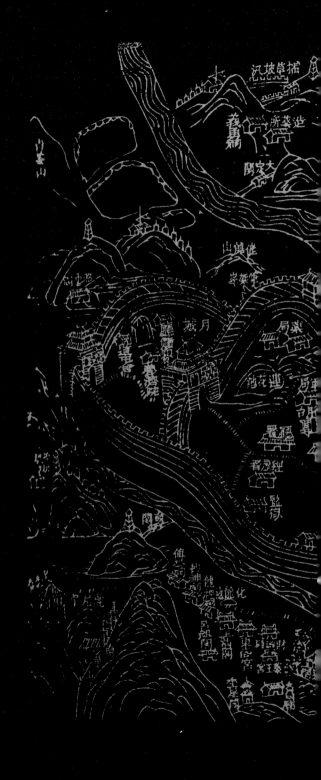

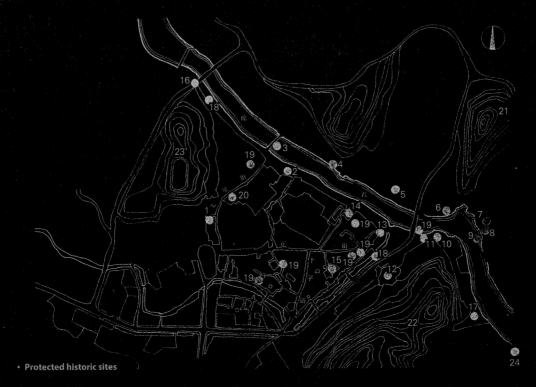

- **Protected historic sites**

0-3 Distribution of Historic Sites in the Ancient Town of Fenghuang

1. Chaoyang Palace 2. Noth Gate 3. Old Beam Bridge(Guliang Bridge) 4. West Pass 5. Ancestral Hall of Family Tian 6. East Pass 7. Wanshou Palace 8. Xiachang Pavilion, Zipingguan Pass 9. Wanmin Pagoda 10. Zhunti Nunnery 11. Huilong Pavilion 12. Three Kings Temple 13. East Gate 14. Ancestral Hall of Family Yang 15. Temple of Town God 16. Nanhua Gate 17. Zhicheng Pass 18. Ancient Well 19. Dwellings 20. Temple of Confucius 21. Qifeng Mountain 22. Guanjing Mountain 23. Bijia Mountain 24. Graveyard of Shen Congwen

I. Ancient Town of Fenghuang

1-1 Bird's-Eye View of the Ancient Town of Fenghuang
Standing on the top of the Qinlong Mountain, one can have a bird's-eye view of the whole town. Nearby are the ancient architectural complex of the Wanshou Palace, Xiachang Pavilion, and Hongqiao Bridge across the Tuojiang River which runs through the northeast of the town. In the distance, the Bijia Mountain is in the northwest. The outline of the town is like a triangle.

The Ancient Town of Fenghuang, lies in the middle reach of Tuojiang River, southeast of the Fenghuang County. With beautiful mountains, lush green woods and clear water, the town is elegant and charming. According to a local folklore, a long time ago, Chinese parasol trees planted there by a merciful immortal attracted phoenix birds, which turned this barren land into a beautiful one. Later, the town got its name from the bird symbolizing auspiciousness and happiness.

According to *Fenghuang Prefecture Records*, "Zhengan", is the old name of Fenghuang Town, which is originated from two places' name, Ganzi in northeast and Zhenxi in northwest of the town. The construction of the town wall began in the Tang Dynasty and was consolidated in the Ming and Qing dynasties. Brick town wall was started to build in 1556 (the 35th year of Jiajing's reign in the Ming Dynasty), and it was changed into stone in 1715 (the 54th year of Kangxi's reign in the Qing Dynasty). Bijia (pen-holder) Town was added as part of Fenghuang in 1789 (the 54th year of Qianlong's reign

1-2 East Gate Tower
Viewing the East Gate Tower from the Chengqianggen Street, one can find the stone steps leading to the gate tower, on which one can appreciate the beautiful scenery of Hongqiao Bridge and Shawan River. Town walls connecting the towers in the past are replaced by residential houses. Without regular maintainence, the gate tower is damaged severely, but still appears majestic as before.

in the Qing Dynasty). And in 1797 (the 2nd year of Jiaqing's reign in the Qing Dynasty), Xiyue Town was built with two gates, the inner west gate and outer west gate. Till then, the total gates in Fenghuang were increased to five. They were the South Gate, Jinglan Gate; the North Gate, Bihui Gate; the East Gate, Hengsheng Gate; the inner west gate, Fucheng Gate and outer west gate, Shengji Gate. The perimeter of the town walls is over 2 kilometers. In 1941, Xue Yue, commander of the 9th war zone of Kuomintang, ordered all counties were to tear down the town walls, but exceptions were those along the Tuojiang River which are allowed to demolish only the battlements and watchtower owing to flood protection. With the expansion of the town after 1949, the walls along the river were also gradually torn down. Only the East Gate and North Gate, connected by a half of the town walls, have been well preserved to the present, which yet retain the majesty of the ancient town.

The town is dotted with historical buildings. Looking at the surrounding mountains at dawn or dusk, you could see Buddhist convents and temples with gray tiles and walls among green pines and verdant cypresses. Amidst the rising smoke of burning incenses, the sounds of bells and drums reverberate in the mountains.

Following emotional descriptions of the town can be found in *Where I Grew up*, an essay written by contemporary writer, Shen Congwen, who was born there.

"…I would like to write about the town once described in one of my works. Although it is only a rough sketch of the place, it seems that all the scenes there is flowing before my eyes, as if they are touchable there."

" If you take trouble to search on an old map with over two hundred years history, you could find a dot marked "Zhengan" at a remote corner at the border of north Guizhou, east Sichuan, and west Hunan provinces.

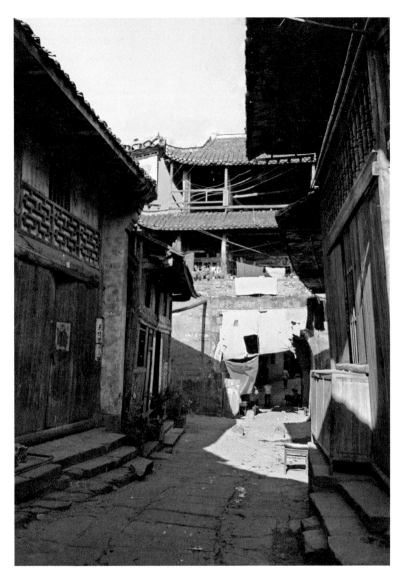

1-3 East Gate Tower at the End of the Cross Street
Observed from the Cross Street, the form and wood material of the East Gate Tower deliver a sense of intimacy. It seems smiling to its neighborhood which is quite different from what one perceives from the embrasures on the other side of the stone walls.

Similar to other dots, that dot stands for a town which can accommodate three or five thousand people. As we all know, the existence of a town and its rise or decline depend mostly on transportation, goods or materials, or trade. But, "Zhengan" is an exception. Its existence is of a meaning intangible. As the center, the round town is built by huge rough stones. Out of the lonely town at the remote border were about 500 forts and 200 barracks. The forts, heaped by large stones, stood on the top of mountains and wound their way to the mountain ranges'while the barracks were well organized along post roads. All those forts and barracks were built at a certain interval according to thoughtful plans about 180 years ago and helped to put down the rebellion of Miao people. All the blood-stained official roads and forts witnessed the tyranny and rebellions in the Qing Dynasty. But, now, all has gone. Most of the forts are in ruins, while barracks have become residential houses. What's more, majority of the Miao people have been assimilated with the Han. At dusk, standing at a high point of the lonely town surrounded by mountains and gazing at the ruined forts near and far, you still can conjure up a faint picture of the past when bugles, drums and torches raised an alarm…"

1-4 Nanhua Gate

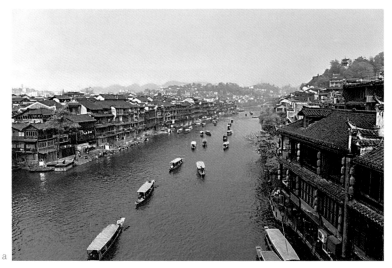

1-5 Scenes I Along the Tuojiang River
a. Boats on the Tuojiang River
b. Hongqiao Bridge and Riverside Houses

"Tourists or businessmen who ever went up the Yuan River by boats and now have a plan to Guizhou Province or Sichuan Province by land without passing by the ancient Yelang Kingdom, Yongshun County and Longshan County, must know that 'Zhengan' provides a good place to have a rest. ... Soldiers are kind and friendly to others as ordinary civilians. Farmers are brave and placid, and all of them respect the God and abide by laws. Peddlers, on their own, carry their empress gauze and other goods, heading for villages in the remote mountains to exchange needed goods and make due profits. The supreme regional governor is the God, secondly the officers, then the village head and the sorcery presenter of God. People there all emphasize self-respect, belief in the God, law abidance and reverence for officers. ...People in this town go to Tianwang Temple every year to offer up sacrifice of pigs, sheep, dogs, chicken, fish or other things they have, and then to prey that God bless good harvests, thriving livestock, healthy growth of children and hold the "Rangjie" ceremony (a religious rite) to eliminate disasters and diseases.

Citizens not only shoulder their responsibility to pay the allocated donation assigned by the government, but also provide financial support of their own to priests of temples or people who perform sorcery. Following the traditional rituals, everything runs in a simple and peaceful way. When it is time to be engaged in farming in spring or autumn, usually, senior people will raise money from different households to perform wood puppet opera for the God of Land and Grain. When there is drought, to pray for rain, kids will carry a dog to hang around with willow twigs or twig-dragon in hand. In spring, Chun Guan (officer who takes charge of cultivation, calendar and sacrifice ceremony) in yellow will read the song of cultivation everywhere. At the end of the year, people there would dress their god of Nuo (a god which drives away pestilence) in red, and place it in the main room of their house, while playing drums loudly as thundering. Meanwhile, Miao witches in bloody red will blow the silver-engraved ox

1-6 Scenes II Along the Tuojiang River (*photo by Zhang Junrong*)
 a. Wuqiao Bridge in the Distance
 b. Wuqiao Bridge
 c. Night View of the Tuojiang River

horn and dance with a copper broadsword in hand to entertain their God. Most residents are the garrison soldiers assigned to defense the land previously. While people from Jiangxi Province would sell piece goods here, Fujian people cigarettes, and Cantonese people medicine. Local upper class is a union of a few scholars and some military officers who join hands in politics and marriages. They implement a stable and conservative policy and govern the town for a long time. What is more, most of the private lands are in their hands. Actually, the upper class was originated from those garrison soldiers there. Tung trees and Chinese fir trees are thriving in the hillsides outside the town and mushrooms can be found every where in the pinewood. Cinnabar hides in the pits while saltpetre is in caves. There will never lack brave, loyal and ambitious soldiers as well as tender, diligent and domestic women. Here, smell of delicious foods always floats out from the kitchens of soldiers, while beautiful voices are coming out from the people who is hewing or chopping."

Reading the graceful words written by Shen Congwen, it seems that we had been back to the remote ancient times and enjoy the pristine style and features of the town.

II. A Town in Landscape

The ancient town of Fenghuang enjoys both fantastic natural environment and wonderful workmanship.

Surrounded by rolling mountains, the town stretches along the Tuojiang River which makes it more delicate and enchanting. In the south of the town, there is the Nanhua Mountain Forest Park where verdant trees cover the hilly valley, rolling like green waves. In the east, there is the Dongling Mountain in which Yinghui Park can be found.

High and sharp, bathing in mysterious morning mist, Dongling provides you a wonderful feast of view with the charming sunrise. A small pavilion stands on the peak of the Dongling where one can enjoy a perfect view, with all the mountains below coming to the eyes at one moment and disappearing in the sea of clouds below in a twinkling. From here, the town seems to immerse in the purple clouds and golden flame. While looking from afar, the town is as bright as a diamond inset on the belt of the Tuojiang River, sparking among the green mountains. A mountain shaped in a pen holder is located in the town, which add it a taste of classic elegance. Ancient buildings with white walls and gray roof tiles, crisscrossed streets and lush plants interweave with each other, all of which make the town more vibrant.

2-1 Scenery Along the Tuojiang River
In the northeast of the town, the majestic North Gate Tower is facing to the Tuojiang River, which is surrounded by low houses on both sides. The ups and downs draw a distinctive outline for the old town. And the verdant trees along Tuojiang River as well as the mountains rolling in the distance paint a landscape scroll.

2-2 Old Wharf and Guliang Bridge Outside the North Gate

The vast space between the North Gate and the river is where the old wharf lay in the past. Now it is preserved and restored for people to take rest and enjoy the scenery.

Another attraction is Magnificent Peaks, which stand on a plain, and spring up like a bamboo shoot with a height of thousand feet into the sky. There are stone steps winding along in the shadow of trees and flowers on the mountain sides, to a temple on the top. There, the echo of the toll make one feel like lost in the tranquil fairyland. In the north of the town, stands a mountain called Magpie Hill. On it, there is an old pavilion called "Jingjin Pass" to which the road was paved with red-stone from the foot of the mountain. The town is constructed against the mountains, which, as a background, make the town more charming.

From west to east, the Tuojiang River runs by the city leisurely. Outside the East Gate, 150-meter-long Fenghuang Ancient Bridge arches across Tuojiang River. On the bridge, one can take a good view of the rough wave rushing through the misty mountains; while turning the eyes to the water below, one can enjoy its clarity and clearness in the shallow part. The large rocks above the water, has been washed and eroded year after year, sleek and mottled. From time to time, ripples in the river are rolling and flickering, from far to near. Rows of Diaojiaolou (stilt buildings) on the banks of the river, with the shadow of the blue sky, white clouds and green trees in the water, draw a fantastic picture. At dawn and sunset, wisps of blue

2-3 New Residences Among Green Mountains and Clear River
Houses along the Tuojiang River are newly built in beautiful environment where traces of history can hardly be found.

2-4 New Look of Tuojiang River
Along the Tuojiang River, there were stilted houses stretching endlessly in the past. With the advance of the time, many of them have been transformed into new residences. However, their horse-head walls and out-sticking corridors still remind us of the traditional houses there in the old days.

2-5 Shawan Scenery
Northeast of the ancient town of Fenghuang takes the clean Tuojiang River as its boundary. The river turns at Shawan and flows to southeast. Three of eight attractions in the ancient town are nearby, namely,"Bridges on Moonlit Streams", "Light of Fishing Boat on Dragon Pool" and "Rolling Waves of Buddhist Pavilion".

2-6 Light of Fishing Boat on Dragon Pool
As one of the eight attractions in Fenghuang, the water here is deep and swift. On the right is Hongqiao Bridge, while the top of Huilong Pavilion and the main hall of Zhunti Nunnery can be seen on the left.

2-7 Door Passage of Huilong Pavilion and Entrance to Zhunti Nunnery

2-8 River Dwellings
The stilted houses along the river are supported by Chinese fir poles against the rock in the river or against the river bank. The overhanging rooms are the boudoirs of Miao girls, and also the place for them to choose their loved ones. They stand at the open space of the stilted house and sing in an antiphonal style with the young man on the river boat. Songs echo between the river and mountains. It is a unique custom in Miao villages.

smoke float over the leisure town. Afar, on the bank where women do their washing, the laughter, talking and the rhythm of clothes-beaten make full delivery of the comfortable and easy life in the small town.

The history has witnessed the beautiful scenery of this town, among which eight are extraordinary—Sunrise at Dongling, Nanhua's Wooded Hills, the Magnificent Peaks, Rolling Waves of Buddhist Pavilion, Morning Bell in the Temple, Bridges on Moonlit Streams, Light of Fishing Boat on Dragon Pool, Woodcutters' Folk Songs along Mountain Path. Most of them are well preserved today. At dawn, climbing up to the Xiachang Pavilion on the Qinglong Mountain in the east, one can still appreciate the beauty of "Morning sun rising in the east, the light waving around with evening mist lingering, and the town is enveloped in purple clouds." The Huilong Pavilion still erects there, greets the wave in the White Pond and Tuojiang River solemnly.

III. Existing Vernacular Antiquities

Differing from grid-like streets and alleys in other traditional cities and towns in China, the layout of Fenghuang ancient town is unique owing to its particular natural environment, economic development, as well as the historical and cultural tradition of minority and region. Radiating from Cultural Square (former Lotus Pond), the main street connects to streets in all the districts of the town, and most of them reach the banks of Tuojiang River, from which we can see that the town has gradually expanded from the river to the mountainous area. It illustrates that the existence and development of the town depends largely on the river.

Tuojiang River, the cradle of the town, rushing passionately past rocks, is with different depths. In the shallow part, one can wade across in dry seasons. While in its depth, there is a deep pool where dragon boating race takes place every year. In the race, the beats of gongs and drums resonate with the voice of cheers. Dragon-boats rapidly scissor the water like flying arrows, marching forward courageously, which excites the audience and also activates the tranquil ancient town.

The river is about sixty to one hundred meters wide, providing a broad and wide view. From the river, the North Gate Tower and the East Gate Tower on the higher positions seem even more magnificent. On the gate tower, the beauty of the whole town and Tuojiang River can be a panoramic feast for the eyes. Aside the banks, exquisite residences are hidden in trees and bamboo fences which creates a beautiful view of different style that one do not want to miss.

The town takes advantage of the local topographic features, so the buildings and the landscape integrate as a whole naturally, which gives itself a fascinating

3-1 Guliang Bridge / opposite
Chinese fir trunks are banded together as a beam, which is laid on the piers and results in a beam bridge. As a unique scenery of Fenghuang, the beam bridge connects two sides of the Tuojiang River. However, the Chinese fir beam will be taken away when confronted with extraordinary flood in case of being washed away.

3-2 Fisherman on the Tuojiang River
Fishing boats like this are frequently seen drifting on the clear Tuojiang River. As works produced by fisherman in Fenghuang, the slim boat is beautiful and light in weight, which can be carried away by a strong man. The fisherman sitting leisurely in the boat seems to take us back to the ancient times.

charm. Such major buildings as the North Gate Tower and East Gate Tower, as well as the Dacheng Hall of Confucius Temple are the visual focuses of the town. Tourists in any main street can catch them easily. Looking from afar outside the town, these major buildings appear like a winding line, ups and downs, becoming the mark of the town and symbolizing the local culture of the citizens. Three Kings Temple (on the Guanjing Mountain) in the east, Wenchang Pavilion and Temple of Mountain Gaurdian (Zhenshan Temple) (at the foot of Huwei Mountain) in the south occupying different heights and nestling in the green trees, echoing to each other in distance, are also an important part of the whole town.

Inside the town, twisting streets divides residential buildings which separate with each other by walls. Thus, winding streets, narrow lanes, high walls and deep courtyards compose a tranquil painting of living. Overlooking the town, crisscrossing with horse-head walls on the roofs, interweave into a cloud of gray roof - a perfect combination of lines and surfaces, which leaves us a profound impression of human beings' creativity.

People in Xiangxi have a firm belief in *fengshui* or Geomantic Theory. Therefore, the location of the town, the sites for major buildings such as temples and altar, even the range of streets and lanes are decided

3-3 Miao Girl / upper
This is the appearance of Miao girls nowadays whose outfit has been much simpler than that of the past. They tie their hair up beneath a headscarf, and wear collarless, slanting front jacket and bell-bottom pants. Often, pleated or embroidered decorations can be found on the collars, cuff of sleeves and bottom of trousers legs. The striking contrast of white and black makes the Miao costumes quite distinctive.

3-4 Going to Fair / lower
In Xiangxi, people still keep the tradition of going to fair every five days (the fair usually is conducted every five days in Chinese lunar calendar). Observing the crowds in the fair, we can find that few young people wear folk costumes nowadays, and those who still dress themselves in traditional way are middle-aged or older women. For male costumes, the little difference between Han people and Miao people is that the Miao men wear a big scarf around head throughout the year.

3-5 Stilted Houses Along Tuojiang River Viewed from the Hongqiao Bridge
This is a photo that could never be seen in real life because most of the stilted houses have been torn down. In winter, the water level drops, and large stones in the riverbed are exposed in the air, so that people could wade over the river at the shallow section.

by principles of *fengshui*. The complex of Xiachang Pavilion, Wanshou Palace and Wanming Tower at the turning corner of the Tuojiang River, are the eye of Fengshui, behind which lies the so-called dragon vein — Qinglong Mountain. While Qifeng Mountain and Guanjing Mountain on its both sides embrace each other to protect dragon vein. Far opposite is the pen-holder mountain in the west of the town. It was called Table Mountain by local people as it is as flat as a table, which was quite beautiful with dense trees around in the past. According to *fengshui* theory, the advantage of the building complex at the turning corner of the Tuojiang River lies that it maintains the vitality of Shuikou (a term in *fengshui* related to fortune). From the perspective of modern urban designing, the building complex is the view focus along the river, which is in line with rules of urban landscape designing. In the past, there was a lotus pond in the center of the town, which was a symbol of good luck and fortune. It is the destination of all the mains streets in the town. People could take a walk on the stone roads and have a rest on the broad land beside the pond. For each street, there is a stone-made water channel through which water either unites at the pond, or flows from the pond to the moat. The water system of the town is like human vessels, running lastingly and vibrantly.

IV. Places for Farewell — Pavilions

Fenghuang is located in mountainous areas. The precipitous paths and interconnected rivers make it quite difficult to get access to the town. Ancient paths are the only way for people to walk from Fenghuang to nearby towns, such as Suoli (Jishou), Luxi, Mayang, and Pushi. Paved with rubble and slates, the ancient paths, rugged and tough, are difficult to walk through. Such paths are high and steep as those to Huoluping, Liangtouyang and Wuchaohe River. With cliffs and valleys on the roadsides, it is quite dangerous for the heavers to pass. As a result, pavilions are built at the flat area of the cols along the ancient paths, for passers-by to take a rest. There are 32 pavilions that have been recorded in the county annals, which are built out of the funds raised by villagers, and their deeds are written into inscription on the stone tablet as models to encourage the later generations. Pavilions, built along the roads outside the town to different places, are called farewell pavilion. The road to Gaocun and Mayang are relatively flat, and served as the main routes for transporting rice, salt, tung oil and local products. About one kilometer away from Fenghuang, there is a pavilion called "Jieguan Pavilion", which is designed for welcoming and seeing off important officials. In the direction of Suoli, there is the "Leicaopo Pavilion", also called "Jingbian Pass", which was built in 1870 or the ninth year of Tongzhi's reign in the Qing Dynasty. On the way to Shuidatian, we can find the "Lengfeng'ao Pavilion", also called "Yingfeng Pass". It was built in 1890 or the sixteenth year of Guangxu's reign in the Qing Dynasty, and can hold more than 100 people. There is still the "Ludi'ao Pavilion", also called "Qingfeng Pass" on the road to Gaocun. The pavilions are places not only for temporary rest, but also for waving goodbye in tears with those who one is close to. In the past, the Miao villages were isolated from outside, thus people had to go through many risks to make a long journey. The suffering from parting beloved ones was even greater when it came to those who were forced to serve army. Those who had to travel far away bid farewell with sorrow to their families and friends at the pavilion. The family members and friends waited there until their beloved ones vanished out of sight, then they, with tears in their eyes, will plant a willow branch or sow flower seeds beside the pavilion to show their cherish. Therefore, each farewell will add a clump of green shade there, year after year, the pavilions are embraced by flowers and willow trees. However, farewell at the pavilion was still sorrowful. In modern society, with the booming of economy and the development of transportation, traveling in and out of the town is much easier, so the pavilions are no longer a sad place for heartbreaking parting. Now they are full of cheers and laughter of young men and women. People say that customs

change with time, the pavilions once symbolized bleakness and depression have been a place full of romances.

Most of the pavilions built at the cols are single-room or double-room style architecture, and some are corridor-style pavilions of three bays. There is another kind of pavilion called "side pavilion", which is attached to the main building, such as residence on the hillside or river bank. Most pavilions are of beam-pillar structures which are supported by four pillars, with Lintels connecting the pillars, so that caravans, sedan chairs could go through it easily. Benches are tied to the lower part of the pillars, so people can have a rest there and shelter from the rain. These pavilions are seldom decorated or just with some simple folk patterns on the roofs, eaves and beams.

In the 1960s, some of the pavilions were torn down

4-1 Pavilion (loggia)
The western Hunan is mountainous, where people in the past depended heavily on foot when they made any trips. Therefore, pavilions were erected along the paths in the mountains so as to have a rest or get shelter from wind and rain. The long journey and difficult transportation posed a serious challenge for those to get out of the mountains. Some people were lost and some even died away from the hometown. So family members would escort the traveler to the pavilion for farewell, thus the pavilion is also called "farewell pavilion".

4-2 Pavilion (shelter)

for building highway. When the new roads came into use, the pavilions along the ancient paths could not play its due role as before, so most pavilions have been left unused or torn down gradually. Today, passing through the Wuling mountainous area by car or by train, the remaining pavilions, as fragments of the history, would remind us of the remote past.

V. Temples and Ancestral Halls

5-1 Three Kings Temple
The Three Kings Temple, which is facing the ancient town, stands on the Guanjing Mountain in the east. On the square in front of the temple, one can take a bird's-eyes view of the town.

Plenty of nunneries and temples are built in Fenghuang in the past. According to the historical records, there are more than fifty temples inside and outside the ancient town. However, some of them are destroyed with the change of time; only twenty are left to the present, but with damages of varying degrees.

Three Kings Temple Three Kings Temple is located at the foot of Guanjing Mountain in southeast of the town. It was built in 1798 or the third year during Emperor Jiaqing's reign in the Qing Dynasty, which is a place for local people to worship three kings—Sun Yinglong, Sun Yinghu and Sun Yingbao, the eighth decedents of Yang Ye (a famous general of the Northern-Song Dynasty). In the past, there were three sculptures of gods in the temple, respectively in white, red and black colors. There is also a big opera stage there. Apart from its regular feature as a worship place with continuous incense offering by its pilgrims. It attracts local people for leisure, especially on festivals, people will gather here for fantastic operas, shows and folk arts. This exquisite architectural complex is located on cliffs. To get there, it is a tough journey: entering from the East Gate, ascending

hundreds of stone steps, and passing through another two gates, there comes the broad yard in front of the main hall. The main hall is magnificent three-room style architecture, supporting by 28 red pillars. The loggia around has only a wooden palisade as its gate. On the central of the palisade, a big round relievo is embedded of "Double Dragon Snapping at a Pearl". With typical local features, the temple is designed to be built with duo-pitched roofs and two horse-head walls on both gables. At the center of the ridge is decorated a huge stone carving of "Double Dragon Playing with a Ball". Other colorful relievo and openwork carving can be found on the beams, columns and *que-ti* (the triangular bracket installed in the corner between a column and the lintel), which are of various themes, such as "A Pair of Phoenix Greeting the Sun" (indicates propitious sign), "Golden Boy and Jade Girl" (attendants of fairies). On both sides of the main hall lie the wing-houses. Across the

5-2 Main Hall of the Three Kings Temple
With horse-head walls and two sloping roofs, the main hall of the temple is similar to common residences except for its bigger size, and delicate decoration. When its fence gates are closed, a round embossment of "Two Dragons Scrambling the Treasure" appears. In the middle of the ridge is the ornament of the same pattern. Under the eave of the central room, the sparrow brace used is as long as the architrave, on which an embossment, called "Two Phoenixes Facing the Sun" is made. And under the architraves of the side rooms are the embossments-"Golden Boy and Jade Girl". All of these exquisitely carved patterns are colorful and gorgeous.

041

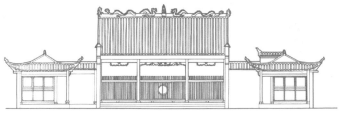

a Elevation

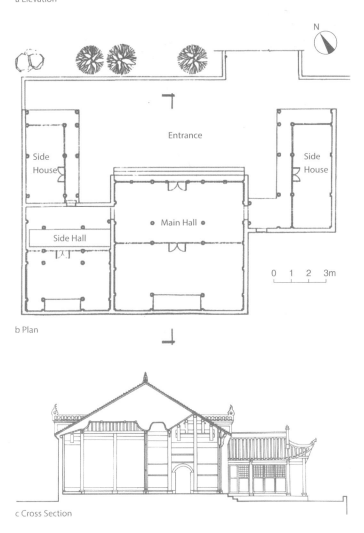

b Plan

c Cross Section

5-3 Plan, Elevation and Cross Section of Three Kings Temple
The temple complex includes the main hall (14 meters long, 15 meters wide and about 10 meters high), both of three side halls on the left and right sides of the main hall, which were restored in 1984.

yard, there are a fish pond and a garden with a variety of fruit trees which offer fruits for all seasons like oranges in autumn, peaches in summer. Except for the destroyed stage, the Three Kings Temple, is well preserved and listed in the major historical and cultural sites under the municipal protection.

Dacheng Hall of Confucius Temple   The Dacheng Hall of Confucius Temple lies at Dengying Street, north of the town. According to *Fenghuang Prefecture Records*, it was first built in 1710 or the 49th year during Emperor Kangxi's reign in the Qing Dynasty, and was expanded and renovated by emperors Yongzheng, Qianlong and Jiaqing, which makes it an architectural complex of a large scale. In front of the hall lies Panchi (a pond in front of an official school), ahead which is Lingxing Gate and Screen Wall. Behind the hall is Chongsheng (respecting the wise) Temple, with Minglun (to know etiquette) Hall in the east. There are other temples around like Minghuan Temple (temple

5-4 Dacheng Hall of Temple of Confucius
The remaining Dacheng Hall is a two-storied wooden structure with a gable-and-hip roof. The roof is high and precipitous and the eave is far stretching outwardly, so it looks dignified and stately, showing the architectural style of State Chu. There are embossments patterned with "Two dragons playing with a pearl" on stone stepway in front of the building and corridor pillars. And the big fan-shaped windows on the side room walls are unique in form.

5-5 Side Roof of Dacheng Hall of Temple of Confucius / upper

5-6 Elevation of Entrance of Chaoyang Palace / lower

The entrance of Chaoyang Palace (i.e. Ancestral Hall of Family Chen) is a symmetrical facade with height-drop horse-head walls and the door opening on a high wall, beside which six embossed landscapes are engraved on the walls. The eaves and cornice are gray and white, and the walls vermilion, which are different from residences around, reflecting a distinctive local flavor.

5-7 Plan and Cross Section of Chaoyang Palace

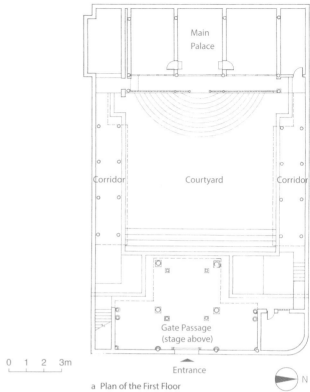

a Plan of the First Floor

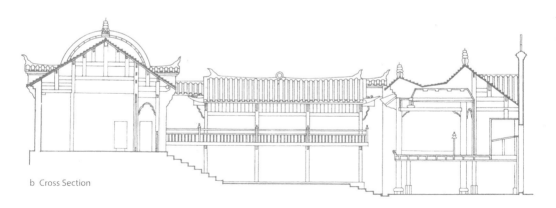

b Cross Section

a

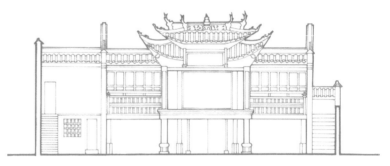

b

5-8 Stage of Chaoyang Palace and Its Elevation
The stage of Chaoyang Palace lies above the entrance, resulting in a rather low gate passage. The stage has double eaves, and there is a gap in the middle of the lower eave. A plaque in the gap writes: "Seeing the Past, Rethinking the Present". Beside the stage is a couplet that goes, "The stage of several feet can be an epitome of a home, a country or the world; the figures of historical influence includes those of the integrity, the evildoer, and the immortals." Behind the stage is a big painting "Fulushou" (means luck, wealth and longevity in Chinese).

to worship respected official), Xiangxian Temple (temple to memorize local talents) and Xingxing Temple (temple for introspection). But some of the temples and halls have been torn down successively since 1949 for the transformation of turning Confucius Temple into No.1 Middle School in Fenghuang. Dacheng Hall, the only part remains, was listed in major historical and cultural sites under the county's protection. The hall, with gable-and-hip roof, is a grand building with three bays. Above the door of the central bay, hangs a plaque with "Dacheng Hall" on it. And the pillars supporting the hall are intertwined by sculptural

a

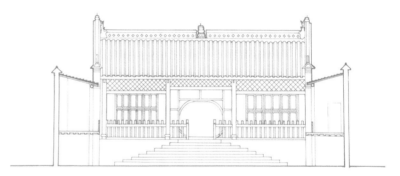

b

5-9 VIP Audience Rooms of Chaoyang Palace and Its Elevation

The VIP audience rooms, facing to the stage, are the best place to watch performance. As the layout of Chaoyang Palace is concerned, the VIP audience rooms are the main hall, in front of which is a semicircular stone stairway. The decoration of the building is simple on the whole, but the oval door surround, as well as the ornate windows on the two sides are especially exquisite. Against the background of the red wall, the windows appears graceful and elegant.

golden dragons, and the beams and architraves are decorated with images of flying dragons, mountains and rivers, as well as flowers and birds. What's more, the eaves of the hall are different from the ordinary ones. At the four corners of the first-tier eaves, four golden dragons are used as decorations, while, for those of the second-tier eaves, four golden phoenixes, all of which wear small copper bells. And a huge portrait of Confucius is hung in the center of the hall.

Chaoyang Palace　Lying at the west side of northern town, Chaoyang Palace, originally called Ancestral Hall of Chen's Family, was built in 1915 or the 4th year of the Republic of China. The exterior wall is over 10 meters in height, with purple stone base and red brick wall. The delicate sculptures of the Aoyu fish (mythical sea turtle) uprising on two ends of the wall are its distinctive feature. The gate in the center of the wall is in the shape of oval, with three Chinese characters "Chao Yang Gong" over it. And an engraved couplet is hung on both sides, stating "An Auspicious bird can fly extremely high in the sky even though it takes off from the little bush; The perceptual audience might be touched by the tower stage even though he views it from the very corner." On the red wall beside it, twelve chic relievos in theme of natural scenery are presented exquisitely. The opera stage is above the gate, resulting in a rather low gateway, and the other side of the gateway is an open space. The square yard paved with slab-stones is less than 400 square meters in area, which can hold hundreds of audiences. The wooden stage, over 10 meters high, is featured with double-tier eaves. And a plaque with Chinese characters of pun is hung between the eaves, reading "Seeing the Past, Rethinking the Present". And in front of the stage, one will find another couplet on both sides, stating The stage of several feet can be an epitome of a home, a nation or the world; the figures of historical influence include those of the integrity, the evildoer, and the immortals. And behind the stage, a huge colored picture is painted on the middle of the wall. Opposite the stage is the

5-10 Part of Temple of Town God / opposite
The architectural complex of the Temple of Town God was greatly destroyed in the past, with only parts preserved. Judging from the remains, one can find they are similar to common residences with two sloping roofs and horse-head walls. The only difference is their bigger size and decoration with simple patterns added to the wood structure under the eaves.

5-11 Ancestral Hall of Family Yang (*photo by Zhang Junrong*) / upper
5-12 Ancient Stage in Ancestral Hall of Family Yang (*photo by Zhang Junrong*) / lower

5-13 Courtyard in Ancestral Hall of Family Yang (*photo by Zhang Junrong*)

main building on a stone base which is 1.2 meters in height. In front of the main building is a big purple stone stair in semicircle shape. The VIP audience rooms (i.e. the main palace) consists of three rooms. The veranda outside is for appreciating opera, while in the rooms one may drink tea and take a rest. Pillars in red and tiles in gray, the building is magnificently beautiful with carved beams and painted rafters. Especially the oval door with openwork surround between the inner hall and the veranda outside is exquisite in craft and stunning in design. In addition, the two storey loggias on both sides of the courtyard are another choice of sites for opera watching.

Town-god's Temple　Located in the southeast of the town, the Town-god's Temple was rebuilt in 1735 or the 13th year of Emperor Yongzheng's reign in the Qing Dynasty, but only part of the buildings survive today.

Ancestral Hall of Yang's Family　Located near the north gate tower, the Ancestral Hall of Yang's Family which is well preserved now, was built in 1836 or the 16th year of Emperor Daoguang's reign in the Qing Dynasty. It was funded by the clansman of the Yang family, at the proposal of Yang Fang, junior guardian of the heir, marquis of brevity, and chief general of Zhengan sub-prefecture. The hall, also named "Hall of *De Xing Ju Kui* (Brilliant Talents)", has a gate in shape of splayed arch, in front of which is a fan-shaped stepway made of purple sandstone. The square gate frame is also made of stone, above which hangs a plaque with four Chinese characters "*De Xing Ju Kui*" in clerical script. And the tall walls of purple sandstone on both sides are decorated symmetrically by over 20 relievos, which, in themes of folk custom scenes, are fresh and brisk. Across the gate is the courtyard, in front of which is the main hall. On both sides of the hall are two storyed wing-houses whose exterior walls are also of purple stone. This is the place where people gathered for a festival celebration in the past, like Tomb-sweeping Day gathering in spring, the Summer Solstice gathering in summer, and the Dead Spirit Festival gathering in autumn, etc. On these occasions, people would sacrifice, entertain and hold rituals here by firing salute, playing music according to their folk customs.

The ancient buildings which survive in the history include: Wenchang Pavilion, Huilong Pavilion, Zhunti Nunnery, Wanshou Palace, Xiachang Pavilion, Wanming Tower. While many more have ruined in the ash of time, such as Qifeng Temple, Dongyue Temple, Yuhuang Pavilion, Jiangjun Temple, Temple of Fire God, Temple of Lord Guan, Temple of Mountain Gaurdian, Zhuge Liang Memorial Hall, Lv Zu Temple, Altar of Land and Grain, Altar of the God of Agriculture, Temple of Dragon King, Yaowang Temple, Guanyin Pavilion, Memorial Temple of Filial Piety, Zhongyong Temple, Fubo Temple, Fugong Temple, Baoguo Temple, Shuifu Temple and so on.

VI. Stone Paved Ancient Streets

6-1 Cross Street I
The Cross Street is a traditional commercial street in Fenghuang, connecting to the Main Street in the north, and the South Gate Tower in the south.

6-2 Cross Street II / opposite
It is a well-preserved section of the cross street, which is listed as the street section under protection in the town.

Street is the sketch of the urban landscape. Walking through streets, one can also catch a glimpse of local folk customs and life style. As a place for local people to communicate, work, rest and entertain, the narrow stone streets in Fenghuang are the best self-portrayal of the town, full of life and local flavors. The crisscrossing streets form a net in the town in which two of them vertically intersect in the center. They are T-shaped Street and Wenxing Street which are the backbone and typical representative of the town road net. They both are traditional commercial streets. One of them is called Main Street by the local people, which extends to the east from Cultural Square, then turns to the northeast and ends at the East Gate Tower. Another street's northern part intersects with the Main Street into a T-shape. Walking along the street to south, one can get to the South Gate, and it is also called Cross Street. The Main Street has always been the business center with flocks of merchants and

6-3 Main Street

The Main Street, a traditional commercial street in Fenghuang, runs through the town from east to west. In the east is the East Gate Tower, and in the west is the Cultural Square. The street is narrow with numerous stores and shops standing on both sides.

6-4 Side Street

Streets and alleys in The ancient town of Fenghuang are narrow but with comfortable ambience. Houses on each side of the streets have their eaves staggering with each other, leaving only a narrow and twisted stripe of sky to watch.

rows upon rows of stores and workshops. The gates of all the shops along the street are street-oriented, with dizzying colorful signboards hanging up either in the middle of the street or beside the shop gate. Though the street is rather narrow, shoppers do not feel constrained and depressed, for there are no walls between shops and streets. Some shop owners would like to set the canopy inclined to the street center, which allows the shoppers to catch only a thread of blue sky and white clouds. Looking up, the uneven eaves on both sides like slightly twisted horizon lines stretch into distance, which makes one feel that there is still a long way to go. The street is paved with irregular rectangular stone slabs, and the intersections between lines of different lengths and plane of various sizes compose a rolling pattern with strong rhythm, on which you may feel like walking into a funny kaleidoscope. However, some small residential streets present another scene with narrow lanes and

6-5 Hejie Street
The Hejie Street is a small street outside the East Gate Tower, whose left is the Tuojiang River. Along the street are shops and stores and behind the shops are dwellings. The small street, paved with stones, is comfortable in size. And the eaves on both sides narrow the sky into a line.

6-6 Bridge Arch Connecting the Hejie Street / opposite
Hejie Street is close to the bank of Tuojiang River, a low-lying area, 4 meters lower than the bridge floor of the Hongqiao Bridge. It is the arch for Hejie Street to run through the bridge approach of Hongqiao Bridge, which plays a role in restricting space as the town gates.

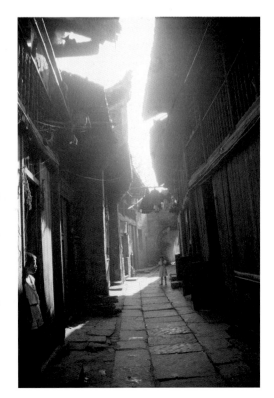

high walls, graceful and tranquil. The towering horse-head walls decorated with white patterns in different shapes, separate rows of courtyards and houses. Those identifiable symbols on the eaves will bring you lots of fun.

Wenxing Street, with a length of 100 meters, is a representative of the residential streets. There are more than 20 households in the street with a slight turning in the middle. Their gates are splayed inward, sparing more room to the street which make the building modest and gentle. Smooth and flat, the Wenxing Street is the best stone street in the town, second to none. The underground drainage channels are wide and deep. Walking in the street in geta during rainy days, the clear pounding sound in the quite place will please one's ear. Behind the street lies Dacheng Hall of Confucius Temple, inside which there were once two giant old osmanthus trees on two sides of the lotus pond. In summer and autumn, the fragrance would overflow the whole street. Among the twenty

6-7 Chengqianggen Street
Chengqianggen Street is a one-side street facing the ancient town wall. There are high stone stepways before the entrances of most dwellings. (*photo by Zhang Junrong*)

6-8 Wenxing Street
Wenxing Street is a famous alley in the ancient town of Fenghuang. The former residence of Xiong Xiling who once was the prime minister of the Republic of China is just on this alley. (*photo by Zhang Junrong*)

household in the street, two of them run manual workshops, one of which is Liu's Dying Shop. Dealing with the color of white and blue with consummate skill, the works made by the craftsman in Liu's have been exported far across seas to America and Europe. The other workshop is Long's Silversmith Shop where workers use fine silver thread to make exquisite silver jewelry in shape of dragon and phoenix, which are indispensable decorations for Miao women.

In the long history, Wenxing Street, tranquil and serene, seems to be telling a story with no end. Children grow up gradually, learn more about the outside world, and build their family at the due age. But most people live the life of their parents, with no more choice, generation by generation. Fortunately, a few of them had the chance to pursue their study and different life out of the old town. Some of them are quite successful in certain fields, such as Xiong Xiling, the first prime minister of the Republic of China, Huang Yongyu, the well-known contemporary painter. They both were born in this ordinary stone street. Therefore, it is believed that Wenxing Street is the home of Wenquxing (a god in charge of literacy in Chinese mythology).

VII. Scenic Spots Along Tuojiang River

7-1 East Gate Tower and East Gate Outer Street
Viewed from the Dongmenwai Street, the east gate tower is simple and plain in appearance. On the stone walls under the gable-and-hip roof, small square embrasures are arranged orderly, as if everything around is still under their control.

The scenic spots and attractions along Tuojiang River have evolved with the passage of time. Based on the modern tourism landscape theory, a complete tourism network has formed in Fenghuang with the improvement of scenic spots and tourism routes. For example, if one tours the town from the East Gate, he can stop at Yang Family's Ancestral Hall on the way to the North Gate. There he can go to the West Pass on the north bank of Tuojiang River through the Old Wharf, during which he can visit the Jumping Rock scenic spot. If he wants to go east forward, there are the East Pass, Zipingguan Pass, Wanshou Palace, Xiachang Pavilion and Wanming Tower along this tourism line. Another route from the East Gate is traveling along the Tuojiang River to the east, and Hongqiao Bridge, Huilong Pavilion, Zhunti Nunnery and Zhichengguan Pass are on this line. At Hongqiao Bridge (also called Wohong Bridge), one can enjoy the sight of stilted houses along Tuojiang River. Turning southeast at Zhichengguan Pass for about one hilometen is Tingtao Mountain where lies the tomb of Shen Congwen, a great modern literature master. One can also follow another route from the East Gate, walking through the Main Street to the Wenxing Street which leads to the North Gate, and then appreciate the vernacular landscape along the Tuojiang River.

1. The Old Gate Towers and the Pass Gates

The East Gate Tower, which is also named as Shengheng Gate Tower, is a tower by Tuojiang River. It is also the east end of Main Street Scene. The tower is a wooden structure building with double-eaves and a gable-and-hip roof. There are neat embrasures on the city walls facing the outside, which appear magnificent. While on the walls towards the town, beams and columns, doors and windows add elegant flavor to the ancient town. On the tower, fort barbettes are built. The tower base which is laid with purple sandstone is solid and strong. In old days, the East Gate was the only access for people, horse and gharries to get into and out of the town. Therefore, in hot summer or rainy days, it would shelter people from burning sun or rain, and vendors were attracted to sell goods there. The local old would like to chat over tea and children were provided lots of fun here. Nowadays, the East Gate Tower has been listed as a key cultural relic site under county level protection.

At the south bank of Tuojiang River, *the North Gate Tower* (also named Bihui Gate Tower), shares the same architecture style with the East Gate Tower. The gate passage is 7 meters deep, with the gate 4 meters tall and 3.4 meters wide. There is a semi-circular barbican outside the gate, whose gate opens to the west. The building of barbican emphasizes the great importance of the gate tower in defense. Out of the barbican gate, walking along the stone steps, one can get to the Old Wharf beside the river.

7-2a North Gate Tower and the Town Wall
The front of the North Gate Tower is like a castle while the side of it is of tower style. The design is quite different, but the transition is natural. The town wall nearby has been renovated in recent years.

7-2b North Gate Tower and the Renovated Town Wall

065

7-3 Zipingguan Pass Gate
Zipingguan Pass gate is the entrance to Fenghuang for people from Mayang County in southeast. It is next to the Wanshou Palace architectural complex. In the past, it not only served as a resting area for the laborious travelers on the roads, but also played an important role in the defense system of Fenghuang.

In recent years, the wall from the North Gate Tower to Hongyanjing in the west has been repaired, which makes the scenery around the North Gate Tower more integrate and attractive.

The Pass Gates In old days, roads and paths were the only means that build up transportation and communications between the ancient town of Fenghuang and the outside, pass gates were built at busy sections of the roads and paths near the town. For example, on the way to Jishou County, there is Liwan Pass Gate, on the way to Mayang County, there is Zhichengguan Pass Gate. These pass gates not only served as a resting place to shelter the laborious travelers on the roads from wind and rain, but also play an important role in the defense system of Fenghuang. The pass gates on the north shore of Tuojiang River include the West Pass Gate and the East Pass Gate and Zipingguan Pass Gate. The former two were the entrances to drill grounds in the

a

b

c

d

7-4 Wanshou Palace

Wanshou Palace, also known as Jiangxi Guildhall. Located in Shawan of Fenghuang, the palace complex is close to the Tuojiang River, from where the landmark of the ancient town of Fenghuang Hongqiao Gallery Bridge could be observed. On its left is the Wanming Pagoda. It is a place mostly concentrated with beautiful scenes and ancient buildings in Fenghuang.

a. Entrance to Wanshou Palace
b. Wanshou Palace
c. Ancient Stage in Wanshou Palace
d. Xiachang Pavilion and Wanming Pagoda Near Wanshou Palace

past, and the latter was the pass gate to the town from north shore of the river in the eastward direction. Nowadays, these rough pass gates have lost their function in defense, and become historical sites with local feature.

2. The Ancient Architectural Complex of Wanshou Palace

The architecture complex of Wanshou Palace, Xiachang Pavilion and Wanming Tower is a major scenic spot along Tuojiang River. Wanshou Palace, also known as Jiangxi Guildhall, was built in the late Ming Dynasty and the early Qing Dynasty when business of Fenghuang began to prosper and thrive. Because most merchants at that time were from Jiangxi Province, the guildhall was completed in 1755 or the 20th year of Qianlong's reign in the Qing Dynasty. The guildhall has over 20 rooms and halls in total, among which the opera stage is a building over 10 meters tall. The main hall opposite the stage is 28 meters long and 15 meters wide, grand and magnificent. Over the main entrance, the plaque is inscribed with four Chinese characters "*Tie Zhu Gong Chong*". Beside, there are two detached side halls with exquisite decoration. Xiachang Pavilion, a hexagonal building with three-tier eaves, is at the southeast corner of Wanshou Palace. It is 20 meters high and the inside diameter of the first floor is 9 meters. Above the main entrance, the plaque bears four Chinese characters "*Lan Sheng Shu Huai*", and there are couplets on the main pillars, one of them goes "Stepping on the great pavilion, overlooking the calm Tuojiang River, clear and clean water rippling the nostalgia into the waves of lakes and seas, long lingering around Taiwan Island; Staring at the border town, concentrating the vigor of sword and brightness of pearl, bearing national spirit in the tranquil forest, riding on the battlefield with

7-5 Xiachang Pavilion / opposite
Xiachang Pavilion, straight and graceful, is the highest building among the architectural complex in Shawan. Standing on its top, one can have a great sight of "Sunrise at Dongling", "Nanhua's Wooded Hills and the "Magnificent Peaks" in the distance, and appreciate the beauty of "Rolling Waves of Buddhist Pavilion", "Bridges on Moonlit Streams" and "Light of Fishing Boat on Dragon Pool" nearby.

magnificence of Wuling Mountain." The eave corners of each floor are decorated with different patterns, which are respectively "Carp Leaping into Dragon's Gate", "Golden Phoenix Spreading Wings" and "Goldfish Swimming in the Sky". Under the eave corners hang copper bells. When wind blows, beautiful tinkling sound can be heard in the sky. Wanming Tower is located to the south of Xiachang Pavilion. The elegant tower is a 22.98 meters tall hexagonal architecture. The diameter of its first floor is 4.5 meters, and that of each floor is 0.30 meter less from the second storey upward. It is built on the base of the ancient "Paper and Words Furnace" which was destroyed in the Cultural Revolution. In 1985, Huang Yongyu, a well-known painter proposed to rebuild the tower, which started in the March of the next year and completed in October of 1988. Because the funds to build the tower were raised by the public, Huang named it as Wanming Tower.

Beam Bridge, Jumping Rocks   Living in such hilly areas with many rivers and creeks which have no plenty of water except the rainy season, people can either wade across or walk on the stone blocks above the water to the other side of the river. Beam bridges are only built on the section where the water is wider or deeper. Wedge-shaped stone piers are placed uniformly in the river to weaken the water potential and stone beams or strapped fir woods are set on the stone piers for people to cross the river. Simple in structure and light in weight, the kind of bridge is safer and more acceptable by the local people for it is not far away above the river surface. The North Gate faces the Tuojiang River, and the beam bridge was the only way for people to enter Fenghuang in the past. In 1980, at the upstream of the beam bridge, a highway bridge was built, which is called Fenghuang Bridge, it is a portal to the town from Jishou. Nowadays, there are other accesses to the town, besides the beam bridge and the North Gate, so they become an interesting part of the town scenery. Though more modern bridges appear on the Tuojiang River, people still prefer to walk through the beam bridge to appreciate the beautiful sights along the river.

Hongqiao Bridge   Originally known as Crouching Rainbow Bridge or the Bridge of Phoenix, it was firstly built in 1670 or the 9th year of Kangxi's reign in the Qing Dynasty, and rebuilt in 1914 or the 3rd year of the Republic of China. Then it is renamed as Hongqiao Bridge. It lies outside of the East Gate, about 112 meters long, 8 meters wide and 9.8 meters high. It is a grand three-arch bridge across the Tuojiang River resembling a beautiful rainbow, therefore it is listed as one of the eight attractions of Fenghuang, known as "Bridge on Moonlit Streams". In ancient times, people once composed poem as:
Plain hills rolling in calm wind,
Bright moon hanging in the sky,
Light mirroring in the water,
Mist covering the village near,

7-6 A Glimpse from Hongqiao Bridge

a

b

c

7-7 a-c Hongqiao Bridge
The ancient Hongqiao Bridge is a stone bridge with three arches. It used to be a gallery bridge, but the gallery was removed because of the construction of a highway in 1955. However it was restored in 1999.
a. Full View of Hongqiao Bridge
b. Entrance to the Gallery (*photo by Zhang Junrong*)
c. Inside the Gallery (*photo by Zhang Junrong*)

Mountains kissing the clouds far,
World becoming pure land.
Wandering on the bridge, back and forth,
Just like strolling in the fairyland.

One can enjoy the sight of "Light of Fishing Boat on Dragon Pool" under the bridge, which is also a best place to appreciate the beauty of the Tuojiang River. In the past, a covered bridge was built on the bridge, beside which were 12 wooden stilted houses served as grocery, food stall and so on. In the middle lay a gallery about 2 meters wide for pedestrians, thus it was actually a unique market street on the bridge. Unfortunately, those covered bridge were removed due to the road construction in 1955. Fortunately the covered bridge was restored in 1999.

The Huilong Pavilion and Zhunti Nunnery　The Huilong Pavilion is located at the south bank of the Tuojiang River, in the southeast of Fenghuang where the geographical condition is harsh. Below the pavilion, there is a red

7-8 Shelter Bridge

Villages in Xiangxi are mostly located along a river or stream and villagers have to cross a river for access to the village, the same case for any roads. Usually a shelter is erected on the bridge to keep people from a bad weather, hence the name. In comparison with that of Dong people in Guangxi, the shelter bridges in Xiangxi are primitive. The shelter is of a column and tie frame with an overhang gable-end roof, or the same structure but with double eaves. As well, some are installed with a pavilion in the middle of the bridge or at the bridge ends, while some of them are attached with local decorations. In recent years, most of the shelter bridges have been removed as a result of construction of highways, leaving only a few survived.

a

b

7-9 Zhunti Nunnery

7-10 Huilong Pavilion Street Near Zhunti Nunnery

The Huilong Pavilion Street is outside the East Gate Tower and to the east of Huilong Pavilion. Compared with those streets inside the town, the street is much narrower so that eaves on both sides almost overlap at certain section. As a result, the street is always in the shadow except for the noon. (*photo by Zhang Junrong*)

sandstone standing out of the river surface. Whenever the river slaps against the red sandstone, great rolling waves will produce huge sound, and it is called "Rolling Waves of Buddhist Pavilion", one of the great attractions in Fenghuang. The pavilion is a gate tower-like building, in the base of which is a passgeway, and overhead the passageway is inscribed three Chinese characters "*Hui Long Ge*". The elegant pavilion on the town tower was rebuilt into fort in the Qing Dynasty in order to put down the Miao uprising. Zhunti Nunnery is located to the south of the pavilion, a stone path outside the pavilion leads to the gate of the nunnery. A few steps away from the path one can walk up a fan-shaped stepway, then reach the spacious court before the main hall. The main hall is 25 meters long, 12 meters wide and 10 meters high with duo-pitched roof and horse-head gables. To the south of the main hall is a side hall whose gable wall is opposite to the gate of the nunnery. Unlike most

7-11 The Graveyard of Shen Congwen
/ upper

The graveyard (circled in the figure) is built according to Mr. Shen's last wish, the gist of which is "Close to nature, simple and natural, no overstatement, no title and no grave". At his graveyard, there is nothing but only a colorful stone backing Tingtao Mountain and facing Tuojiang River. The stone, with a weight of about 6350 kilograms, 1.9 meters in height and 1.5 meters in width, is crude and natural. The inscription on the stone goes, "Be no servile and slavish" and "Be kind and be tolerant" is an ideal portrait of his personality and his life pursuit.

7-12 Tombstone Of Shen Congwen
/ lower

The tombstone is made of a natural multicolored stone. Shen's handwriting is engraved on the front of the stone, below which are retaining walls. The beautiful environment and plain tombstone is a real portrait of Shen's personality.

of the traditional Chinese temple, the main hall and the side hall are asymmetric in layout. On the hill behind the nunnery is a spring, called Hongshan Spring, whose water will never run out throughout a year. Because of the geographic advantage, standing in front of the main hall and looking towards north, one can have an amazing view of "Sunrise at Dongling" and "Magnificent Peaks", two of the well-known attractions in Fenghuang.

The Graveyard of Shen Congwen　　The graveyard of Shen Congwe is located at the Tingtao Mountain, by the side of Nanhua Mountain. It is believed that it is a place with an aura of the heaven and earth because it is surrounded by mountains, cliffs and Tuojiang River. The mountains around are covered with verdant, old plants, and spring water is dripping from the caves, quiet and peaceful. According to Shen's last wish, his tomb is built. But there is only a giant colorful stone like a mushroom without a tomb or stele with inscription. Under the stone is part of Shen's bone ash because most of it has been spread into the Tuojiang River. Besides, the ash is not contained in casket but mixed with soil so as to nourish the earth. The colorful rock is taken from the Tingtao Mountain. It is set before the precipice in a primitive state without being carved or polished, and it looks like a natural part of the precipice, an ideal destination for a literary giant. On the front side of the rock, Shen's handwriting is engraved, which goes as "Follow my mind and you'll see a real myself; follow my mind and you'll understand the true ego of man." which is from the *Autobiography of Shen Congwen*. The back of the rock is engraved with "Be no servile and slavish, his works are shining and glaring; Be kind and tolerant, he is loyal and sincere" which is a lively portrayal of his personality. This excerpt is selected from the elegiac couplet given by Zhang Chonghe, his wife's fourth sister in 1988 when Shen passed away. Zhang was a professor of Yale University, a sinologist and calligrapher. On October 18th 1993, Zhang made a special trip from USA to Fenghuang only for paying homage to Shen and wrote inscription "Fenghuang, as ancient as fascinating, is much more outstanding and elegant because of the great contemporary writer Shen's cemetery, which can be taken as the most beautiful place in the town and can be cherished generation by generation. I am honored and lucky enough to visit his cemetery!"

VIII. Courtyard Dwellings and Stilted Building

There are two major types of traditional residences in the ancient town of Fenghuang. One is the urban courtyard houses distributed in the town, which connected with each other into "*Fang*" ("*Fang*" is the basic residential unit in ancient China, spaced by streets or roads). The other is stilted buildings, the most distinctive architecture in Xiangxi (the western Hunan), distributed at the river banks or in the hills and cliffs around. Traditional stilted houses are mainly of wood structure except for the tiles. Therefore, it is not easy to preserve because of fire risk. In modern society, with the emergence of new building materials and techniques, as well as lack of wood resources, residential dwellings change as a result. Now, the stilted buildings remaining in the town are rare. Fortunately, we can still find their trace on the banks of the Tuojiang River.

1. Courtyard Dwellings

Unlike stilted houses, which were of Xiangxi local residential style, the courtyard dwelling are a building type introduced from outside. But still, unlike the traditional Han people's counterpart in shape of square, the plans of courtyard dwellings in Fenghuang are, in most cases, of irregular geometric shape subject to mountain terrain. Because the arrangement of the streets and lanes do not keep to the direction of due north and due south, the distribution of the houses seems to be more free and irregular. Larger dwellings are often of 2 stories, with a patio at the center, connecting to the halls and rooms on the two floors. In front of the patio lies a passage to the gate, and at the back, an open hall. On each side of the patio, there is an open corridor, where the staircase is built (some are built at the side of the open-hall). Often there is a winding corridor on the second floor surrounding the patio on which stands a pavilion, sheltering the patio from rain. Generally, the pavilion is a freestanding structure without enclosures and is taller than surrounding houses for better sheltering, airing and lighting. A patio roof can also be made by lifting the pillars at the outer eaves, but it makes the patio more enclosed. To make the patio bright, glossy tiles are usually used on the roof. The patio has both outdoor and indoor functions, and it is an essential place for residents' life and work. *Sanheyuan* (A living space composed of buildings on three sides and an enclosing wall) is another compact courtyard, of which the main house and wing houses are often of two stories. The external walls and

8-1 Bird's-Eye View of Former Residence of Shen Congwen / left
Located in the Zhongying Street in the southern part of the town, the residence is a plain courtyard with the main house and gate house both of three bays, as well as two wing houses of two bays. On December 28th, 1902, Mr. Shen Congwen was born here. The former residence of Shen is officially open to the public after renovation, in which his handwritings, manuscripts, relics and portraits are displayed.

8-2 Entrance to the Former Residence of Shen Congwen / right

gables are unparallel with the main house because they have to adapt to the terrain of the street, thus trapezoid courtyards of various size are formed. The gate of the *sanheyuan* courtyard is not in the front, but on its side. It is facing the wing house through the open corridor. For most of the ordinary families, the gate is plain in decoration, with only simple patterns engraved on the stone frames. And over the gate is a single pitch eave, covering a recessed tiny place, where is the space for family members to have a short rest, as well as a shelter against bad weather for passers-by.

8-3 Inner Courtyard of the Former Residence of Shen Congwen

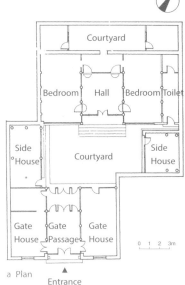

a Plan

8-4 Plan and Cross Section of the Former Residence of Shen Congwen
The residence is special in its dislocation between its front and rear part. There are courtyards in the middle and rear part. It is compact in layout and simple in appearance.

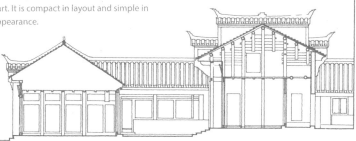

b Cross Section

8-5 Bird's-Eye View of Residence of Family Xiang
The two-story building compound shapes in resemblance of Chinese character "冂" with its main entrance on the east side. The layout of the whole compound is very compact with the bedrooms on the second floor and the hall on the first floor.

8-7 Cross Section of the Residence of Family Xiang
The Xiang's residence, with a compact layout, is a typical example of *sanheyuan* courtyard. The main house and side houses are all of two stories. The courtyard wall and gable walls of the side houses are not parallel with the main house as they have to be alined with the street. Therefore, the room sizes in the side houses are varied.

8-6 Gate of the Residence of Family Xiang
The Xiang's residence is located at a narrow alley, whose gate is a little bit recessed and the side walls splay slightly as a Chinese character "八"(eight). The kind of the gate is common in Xiangxi, which means modesty, concession, and hospitality.

083

8-8 Gate and Porch of the Former Residence of Xiong Xiling

Xiong Xiling was talented since his childhood, reputed as a genius. He stood out successively in the serial examinations, including the national civil examination. Then he left Fenghuang and later became a known stateman, who was once the first prime minister of the Republic of China. In a small alley in Fenghuang, the former residence of Xiong Xiling is quiet, and simple, natural and exquisite in style.

8-9 Courtyard of the Former Residence of Xiong Xiling

8-10 Plan of the Former Residence of Xiong Xiling

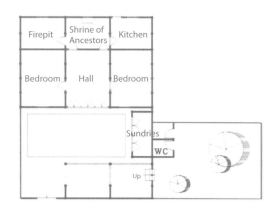

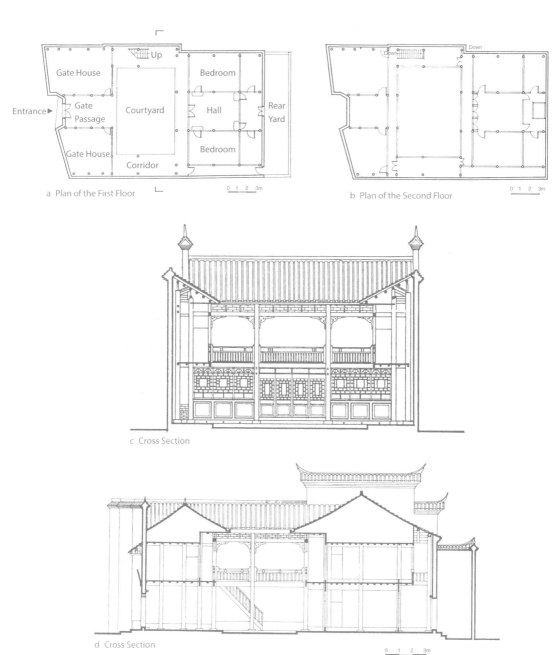

8-11 Plan and Cross Section of the Silver Screen Family

The Silver Screen Family is Zhou Family's residence, where some movies were made, hence its name of "Silver Screen Family". The houses in the front and rear are both of two stories. On one side of the patio, a straight stair leads to the interconnected verandas on the second floor. Therefore, the patio functions as a hub in connecting the front and the rear, and different rooms on both the upper and lower floors. There is a narrow small patio in the rear of the residence, with a passage on one side to the front patio, forming a loop. In the courtyard, trees and flowers are grown.

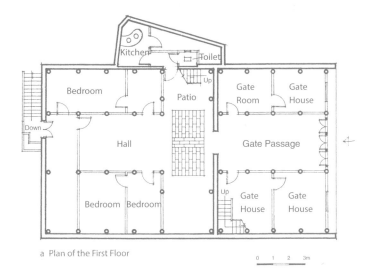

a Plan of the First Floor

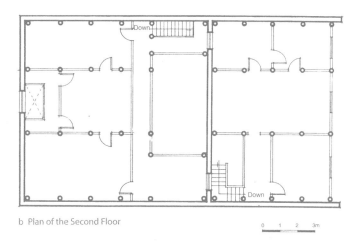

b Plan of the Second Floor

c Cross Section

8-12a~c Plan and Cross Section of Xiong Fanwen's House

In a compact and orderly layout, Xiong Fanwen's house in town, is adjacent to streets on both front and back sides. The patio is small and narrow, with a kitchen and a toilet on one side. On the second floor, veranda surrounds the patio from three sides while stairs are set in the patio on the first floor. But the second floor in the front part is disconnected with the veranda of the patio. Instead, it is an independent section with another stair leading to the second floor.

2. Stilted Building

As a building structure with a long history, the stilted building is believed by some that it has developed from another ancient residential type – nest dwelling. The form of stilted house, which can be divided into two major types: Tiaolang style (with overhanging corridor) and Ganlan style (with stilted pole) vary in accordance with different regions, ethnic groups and topography. The stilted buildings along the Tuojiang River are of Ganlan style which is a buildings on wooden poles, with the upper floors as living spaces. They are tightly bound to rivers and beaches. Usually they are constructed along rivers or beaches in large groups, and the magnificent extension will impress one a lot. Most stilted houses of Ganlan style are buildings of two stories, some three stories in part, with overhanging gable eaves. The horizontal line produced by waist eaves and the waist corridors, and the changes of light and shadow made by belt-shaped windows under the waist corridors form a violent and vivid contrast with the wooden poles below, which makes the stilted houses more attractive. Passages are built at intervals among the stilted houses along the river, with stone steps leading to the piers. A house overhead the passage that is not detached with the stilted houses on the two sides is then a street-stradding building.

8-13 Stilted Building Along the Tuojiang River
These stilted building are on the south bank of the Hongqiao Bridge and the east side of the river. On the other side is the Hejie Street. Endless stilted houses hanging over the water surface have become an attraction of the river. Owing to the scarcity of wood, and the fire risk, only a few of them remains. Therefore, the eight houses are listed as the buildings under protection in the ancient town's planning. In the future, the internal space and living conditions of them will be improved and they may be changed into hotels to accommodate tourists.

Most of the plans of stilted houses are of two or three rooms, in a slightly fan shape, as a natural result of building against the curve of the river banks. The side of the houses along the river is equipped with a corridor where Miao girls can make embroidery or sing love songs with young men on the river boat. The other side serves as the shopping area along the street. The main body of the stilted house is the column and tie frame. To tally with the landform of the river bank, the size of the room is of great flexibility. However, on the average, the rooms are 3 meters in width and 10 meters in depth.

8-14 Supporting System of the Stilted Building / upper
Most of the stilted houses are supported by fir poles, with bamboos and other materials as supplement. The supporting forms vary according to stress distribution, which can be vertical or aslant support.

8-15 Houses along the Moat / lower
The moat circles the ancient town in the south and southeast, and links with the lotus pool in the town center. The moat, with clear and dynamic water flow, meets the Tuojiang River outside the town as well. On its two sides are stilted houses, which provide a fantastic view. However, with the silting-up of the moat, decreasing of the water flow and the enviromental pollution, houses on its sides also gradually decline.

a Elevation

b Section

c Plan

8-16 Elevation, Section and Plan of Stilted Buildings along the Tuojiang River
Stilted buildings along the Tuojiang River are typical examples of waterside stilted buildings in Xiangxi. They are arranged along the banks casually in a fan-shape over the river. Their supporting pillars either lean on the river walls or erect on the rocks in the river. There is no space left between the houses, which stretch out into a well-proportioned landscape. However, many of these houses are either removed or destroyed by flood.

8-17 Stilted Houses Along the Moat / opposite

Chronology of Significant Events

Dynasty or Periods	Reign Titles	Years(CE)	Events
Qing	39th year of Emperor Kangxi's reign	1700	Fenghuang prefecture established.
	49th year of Emperor Kangxi's reign	1710	Confucius Temple built in Dengyin Street within the prefecture.
	54th year of Emperor Kangxi's reign	1715	The Zhen'gan town's wall was changed from brick to stone one. The circumference of the town wall is more than 2000 meters and the town wall reached a height of 5 meters.
	12th year of Emperor Qianlong's reign	1747	Jingxiu Academy established in Dengyin Street for children and young people to study.
	23rd year of Emperor Qianlong's reign	1758	Fenghuang Prefecture Annals complied for the first time by prefecture officials Pan Shu and Yang Shengfang.
	60th year of Emperor Qianlong's reign	1795	In early January of the lunar calendar, the Miao people at the boarder of Hu'nan Province and Guizhou Province started an uprising against the Qing government. Wu Tianfeng, and Wu Longdeng (later defected), two Miao residents of the Fenghuang County, led the local residents to join the uprising.
	1st year of Emperor Jiaqing' reign	1796	In order to suppress the revolt of Miao people, Fu Nai, vice-magistrate of Fenghuang, pursued the policy of "Tun"- to confiscate the weapons held by common people, appoint officials in charge of opening up wasteland by farmer-soldiers, build blockhouses and cordons, open up wasteland and train soldiers.

Dynasty or Periods	Reign Titles	Years(CE)	Events
Qing	3rd year of Emperor Jiaqing' reign	1798	Sanhou Temple (also named the temple of the Heavenly King) was built at the foot of Guanjingshan Mountain outside the East Gate.
	5th year of Emperor Jiaqing' reign	1800	After taking office as vice-magistrate of Fenghuang, Fu Nai repaired border walls of 55 kilometers, and built 848 fortifications, including barbettes, pass gates, blockhouses and cordons from the second year to the fifth year of Jiaqing' reign.
	12th year of Emperor Jiaqing' reign	1807	Setting up 27 free private schools for children of Han people and soldiers as well as 25 free private schools for Miao people.
	4th year of Emperor Daoguang's reign	1824	Completion of the *Fenghuang Prefecture Record* by Sun Junquan and Huang Yuanfu with 10 books and 12 volumes.
	17th year of Emperor Daoguang's reign	1837	Hu-Guang Viceroy Lin Tse-hsu inspected Zhengan, suggesting the Qing government to exempt the the reclaimed land lease of 250,000 kilograms between the 11th year and 14th year of Emperor Daoguang's reign and scrap those damaged farmland about 233 mu or 15.53 hectares.
	13th year of Emperor Tongzhi's reign	1874	The Miao personage Wu Zifa raised money to construct Santan Academy at Beitingao of Deshengying during his governing in the east part of Guizhou Province.

Dynasty or Periods	Reign Titles	Years(CE)	Events
Qing	32nd year of Emperor Guang Xu's reign	1906	Establishing four prefecture middle schools (Fenghuang, Ganzhou, Yongsui and Huangzhou), appointing Tian Xingliu as principal, who studied from Japan Hongwen Normal University.
	3rd year of Emperor Xuantong's reign	1911	The news of Wuchang Uprising came to Fenghuang County. And local personages Tang Shijun and Tian Yingquan tegather with their followers made a response to the event on October 27th in Chinese calendar by establishing the army of restoration, and attacked the county seat but failed with a loss of more than 170 people.
The Republic of China		1912	Miao, Han, Tujia People from four counties (Songtao, Fenghuang, Qianzhou and Yongsui) gathered together to stage an uprising. On January 1st, (November 13, 1911 of the lunar calendar) a regime —Xiangxi military-political government was founded.
		1914	The grave and monument were erected in honor of the revolutionary martyrs; the Craft Training School for Women was founded at Tengjiawan of Ximenpo. Tian Yingbi, who had studied in Japan, was appointed as the schoolmaster; the Chinese Brush Painting School was established at the ancestral hall of Family Tian in Lao Yingshao. Que Hongyuan was appointed as the schoolmaster.
		1917	Tian Yingzhao, a military official of Xiangxi, led the "Gan Troop" eastward to Yuanling, to uphold the constitution in Hu'nan Province.
		1925	A serious disaster hit the land. Xiong Xiling and Mei Lanfang respectively donated 60000 yuan and 40000 yuan to relieve the damage. However, many poor people still were forced to sell their children and starved to death.

Dynasty or Periods	Reign Titles	Years(CE)	Events
The Republic of China		1926	American churches in Changsha assigned two missioners to Fenghuang (one is an American and the other is a British), and set a gospel church at the main street. Since then, Christianity was introduced into the county.
		1929	Yuanling Catholic Church sent Priest, Yang Menglin to Fenghuang County and set a catholic church at the main street. Since then, Catholicism was introduced into the county.
		1937	In July, the Fenghuang party branch of Kuomintang established its official newspaper: Fenghuang People's Newspaper, which mainly published news extracts about the Anti-Japanese War and bulletins of the county government at the beginning.
		1938	In spring, the CPC Hu'nan Working Committee took the opportunity of establishing the Refugee Workhouse for Women & Children by Hu'nan provincial government. The committee sent Bai Yunhua as the secretary, Huang Shaoxiang as committee member to set up the CPC Fenghuang County Committee whose work is under the cover of the Workhouse. In March, the organization started its official work in Fenghuang.
		1941	Xue Yue, the general commander of KMT in Hu'nan ordered to tear down town walls of all counties. The town walls of Fenghuang County were all torn down, except the part along the river. The battlement and defense tower of this part were dismantled, but the main body remained for the purpose of flood control.
The People's Republic of China		1949	Fenghuang County was peacefully liberated and temporary Public Security Committee was established. In 1949, Fenghuang had a total population of 156.3 thousand, among which 81.4 thousand were Miao minority.

References

1. 魏挹澧.湘西风土建筑—巫楚之乡 山鬼故家.武汉：华中科技大学出版社，2010.
2. 沈从文.沈从文文集（第九卷）.长沙：湖南人民出版社，2013.
3. （清）黄应培，等.凤凰厅志（卷一）.长沙：岳麓书院，2011.

The Series of 100 Gems of Chinese Architecture, Chinese Edition
Chief Planners: Zhou Yi, Liu Ciwei, Xu Zhongrong
Chief Editor: Cheng Liyao
Assistant Chief Editors: Wang Xuelin
Editors in Charge: Dong Suhua, Zhang Huizhen, Sun Libo
Technical Editors: Li Jianyun, Zhao Zikuan
Photo Editor: Zhang Zhenguang
Art Editors: Zhao Qing, Kang Yu
Layout Design: Nanjing Hanqingtang Design

The Series of 100 Gems of Chinese Architecture, English Edition
Chief Planners: Shen Yuanqin, Sun Libo, Zhang Huizhen
Editors in Charge: Dong Suhua, Zhang Huizhen
English Text Proofreading: He Guangsen et al.
Technical Editors: Li Jianyun, Zhao Zikuan
Photo Editor: Zhang Zhenguang
Art Editors: Zhao Qing, Kang Yu
Layout Design: Nanjing Hanqingtang Design

The Series of 100 Gems of Chinese Architecture

ANCIENT TOWN OF FENGHUANG
凤凰古城

Text & Photo by Wei Yili
Translated by Chen Zhaojuan / China Translation Corporation

All rights are reserved. No part of this publication may be reproduced, distributed or transmitted in any form or by any means, including photocopying, recording or other electronic or mechanical methods, without the prior written permission of the publisher, except in the case of brief quotations embodied in critical reviews and certain other noncommercial uses permitted by copyright law.

Copyright ©2019 China Architecture & Building Press
Published and Distributed by China Architecture & Building Press
ISBN 978-7-112-23510-0 (24445)
CIP data available on request
www.cabp.com.cn
Printed on acid-free and chlorine-free bleached paper

Printed in China